HARCOURT SCIENCE

VIRGINIA SOL SUPPORT FOR STUDENTS

GRADE 2

Orlando Austin Chicago New York Toronto London San Diego

Visit *The Learning Site!*
www.harcourtschool.com

Reviewers

Sharon Bowers
Kemps Landing Magnet School
Virginia Beach, Virginia

Brenda Dorman
Coleman Place Elementary
Norfolk, Virginia

Coleen Matthews
Coleman Place Elementary
Norfolk, Virginia

Diane C. Tomlinson, Ed. S.
Virginia Science Content Specialist/Codirector
Coalfield Rural Systemic Initiative
Lebanon, Virginia

Requests for permission to make copies of any part of the work should be addressed to School Permissions and Copyrights, Harcourt, Inc., 6277 Sea Harbor Drive, Orlando, Florida 32887-6777. Fax: 407-345-2418.

Printed in the United States of America

ISBN 0-15-338712-2

12 13 14 15 1417 13 12 11 10

4500239835

Contents

Contents *(continued)*

Name ____________________ Date __________

LESSON 1

Vocabulary
characteristics

Classifying Objects

Describing Objects

Remember that when you classify things, you sort them into groups by ways they are alike. Objects may be alike in many ways. They may be the same color, the same size, or the same shape. They may feel the same or have the same smell. These ways to describe objects are called **characteristics,** or properties.

Using More Than One Characteristic

Objects can be classified by more than one characteristic. You may put a set of balls in the same group because they are all the same color and all soft. Shells may be put in the same group because they are all the same shape and all smooth.

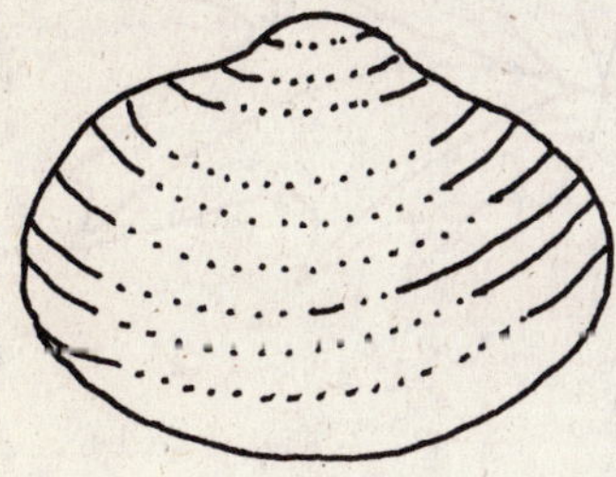

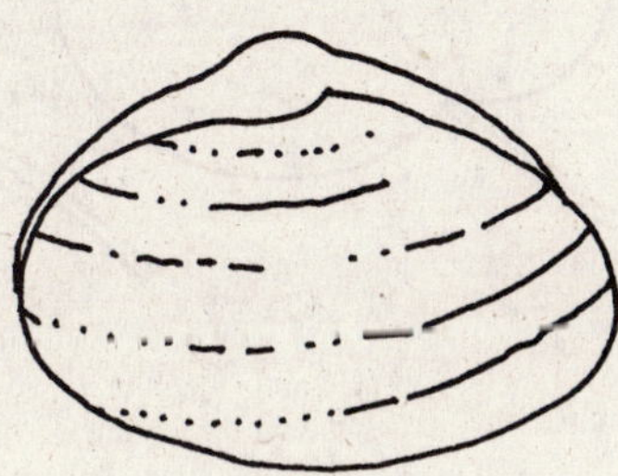

■ **What are two ways these shells are alike?**

Name ______________________________ Date ____________

Activity

Directions Circle all the leaves that have smooth edges.

Color green all the leaves that have rough edges.

What other characteristic can you use to classify the leaves?

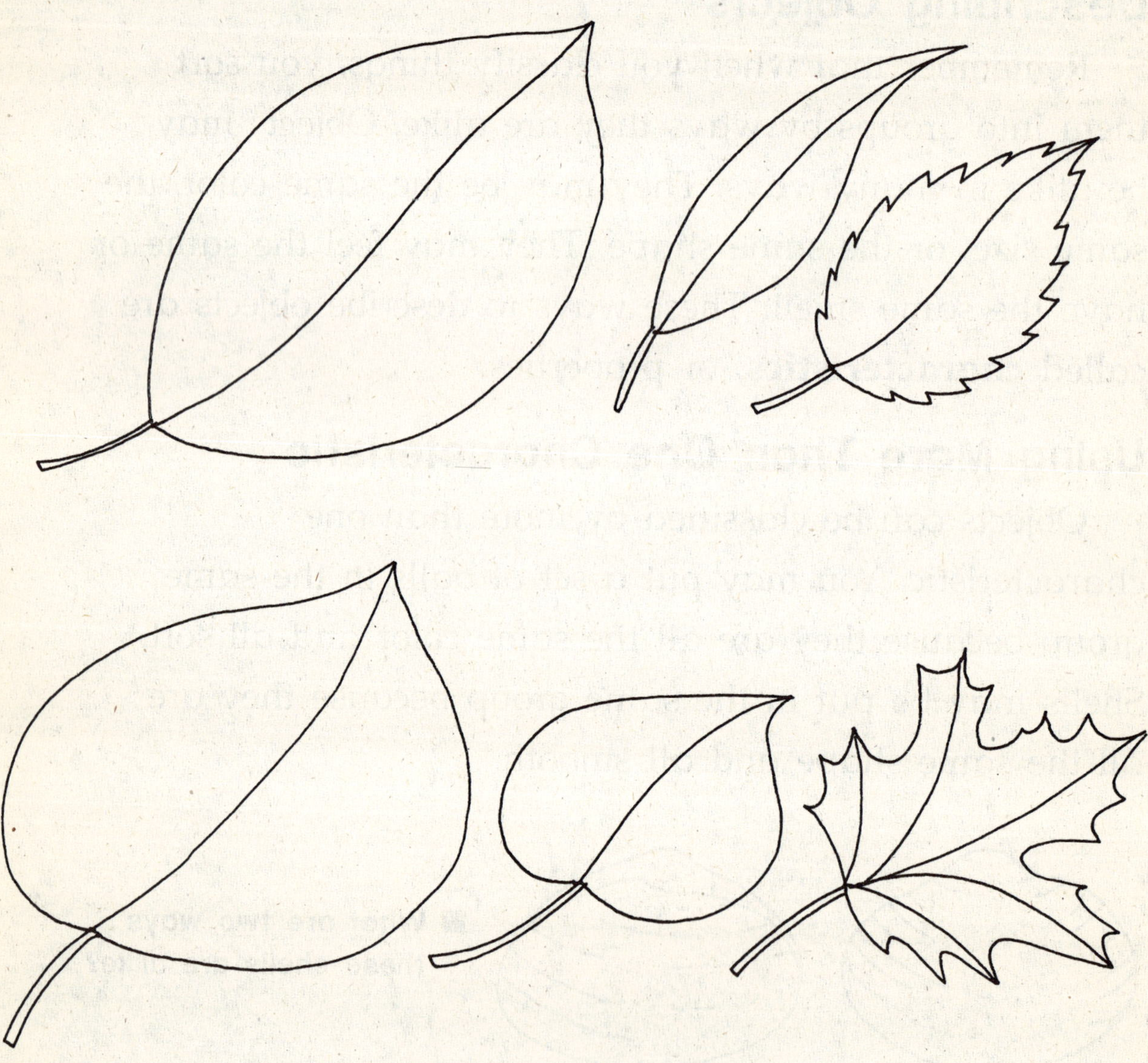

Name ______________________ Date ____________

LESSON 2

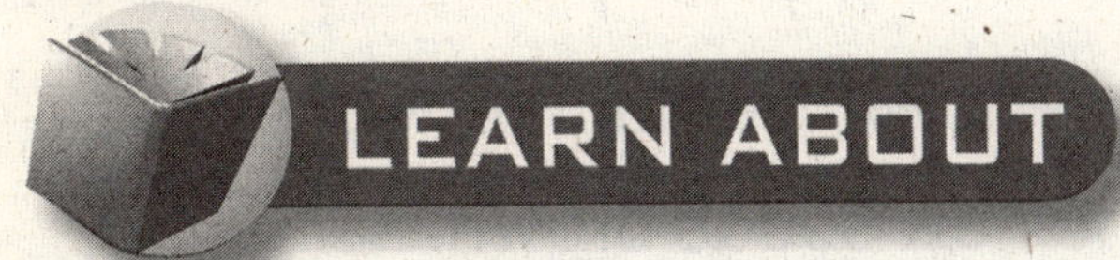

Vocabulary	
life cycle	seedling
buds	

Plant Life Cycles

Living Things

All living things change as they grow. A **life cycle** is the set of changes a living thing goes through.

Plants Have Life Cycles

Most plant life cycles begin with a seed. A tiny plant sprouts from the seed and begins to grow. A new young plant is called a **seedling.** If a plant has flowers, the flowers start as small **buds.** The buds open into flowers. The flowers make seeds. Some plants do not have flowers. Their seeds are made by other plant parts. Sometimes seeds are found in the fruit or the nut a plant makes.

The new seeds fall from the plant. They land in the soil. The life cycle begins again.

Different Life Cycles

Some plants grow slowly. Other plants grow fast. An oak tree grows very slowly, so it has a long life cycle. A bean plant grows fast, so it has a shorter life cycle.

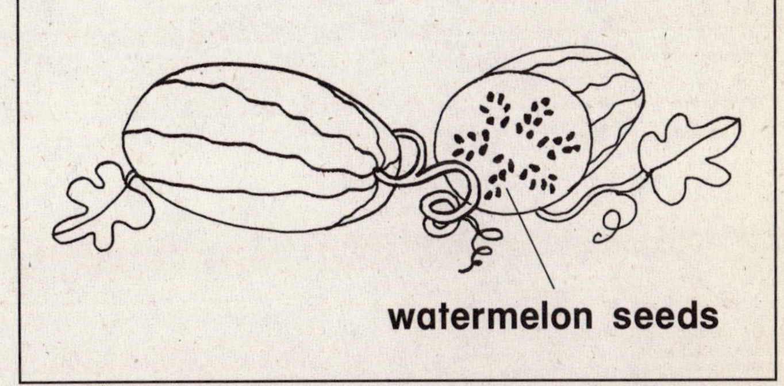

Name ______________________ Date __________

Activity

Directions Number the pictures to show how the plant grows. Color the planted seed black. Color the new seeds brown.

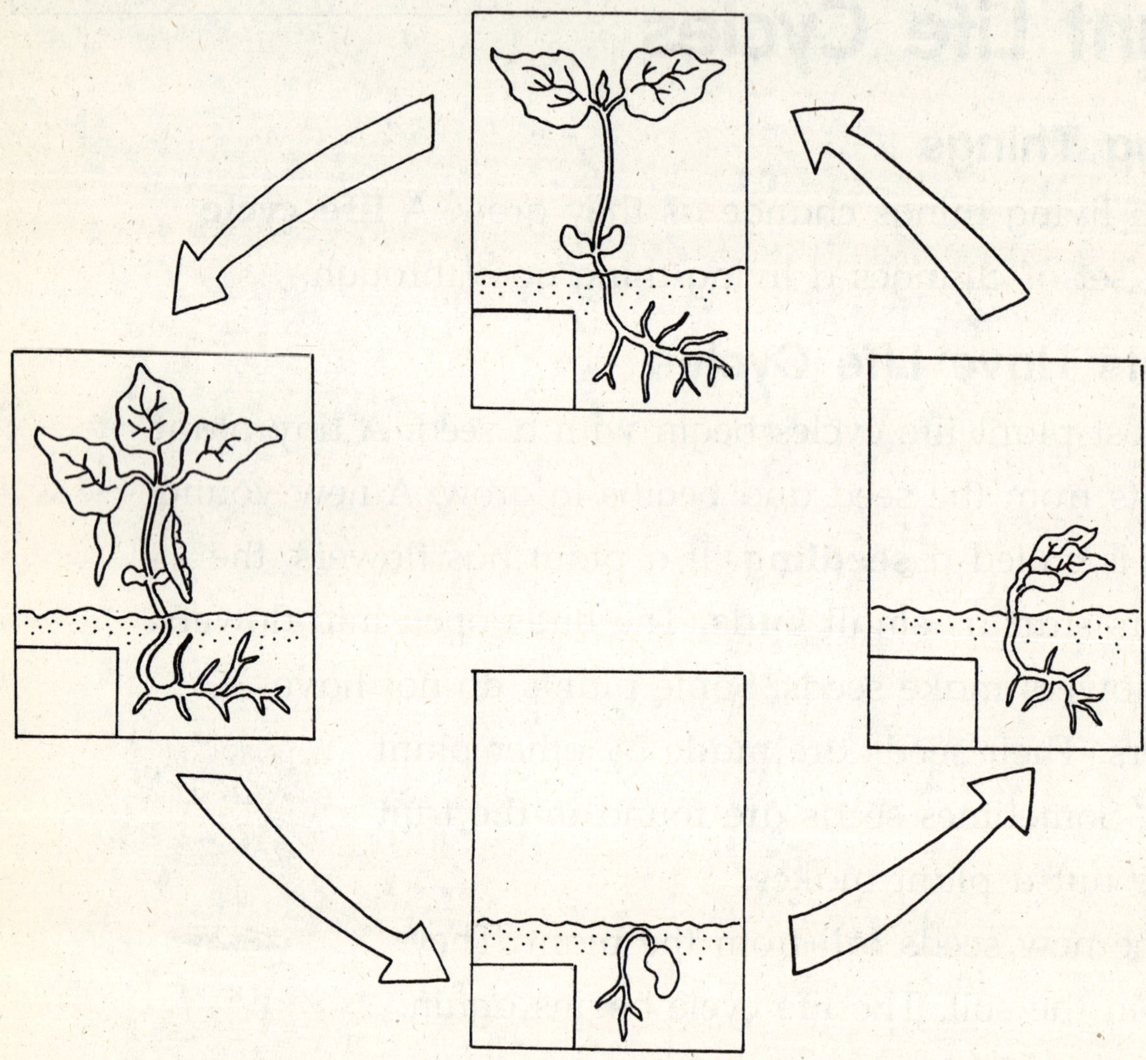

Name ______________________ Date ____________

LESSON 3

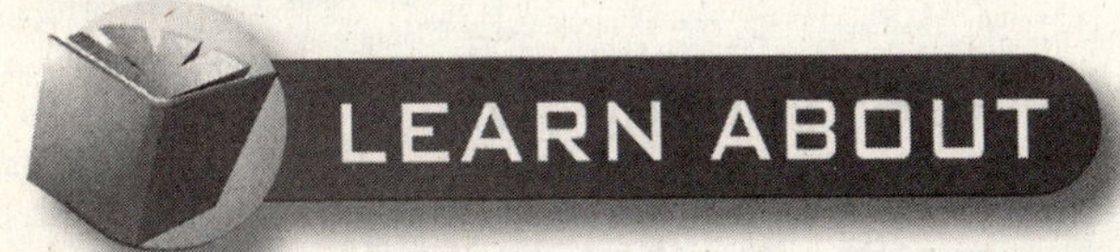

Vocabulary
deciduous

Deciduous Trees

Trees Change Each Season

In summer most trees are covered with leaves. By winter, the leaves of many trees are gone. Why? These trees have leaves that change with the seasons. They are called **deciduous** trees. A good example is Virginia's state tree, the dogwood.

dogwood tree in summer

dogwood tree in fall

Spring and Summer

In spring it is warm and there are more hours of sunlight. Deciduous trees begin to grow new leaves. By summer the trees are covered with healthy, green leaves.

Fall and Winter

Fall brings cooler temperatures and less sunlight. The leaves of deciduous trees change colors and then drop off. By winter their branches are bare.

Name ______________________ Date __________

Activity

Directions Draw and color a deciduous tree in spring. Draw and color a deciduous tree in fall.

Spring	Fall

 SOL 2.7a Use with Unit A, Chapter 1.

Name ______________________ Date __________

LESSON 4

Vocabulary
dormant

Weather Affects Plant Growth

Temperature and Plants

Most plants grow well in spring and summer. Warm temperatures and lots of sunlight help plants make food and grow.

Many plants stop growing in fall and winter. With cold temperatures and less sunlight, they cannot make food and grow. When plants stop growing, they are **dormant.**

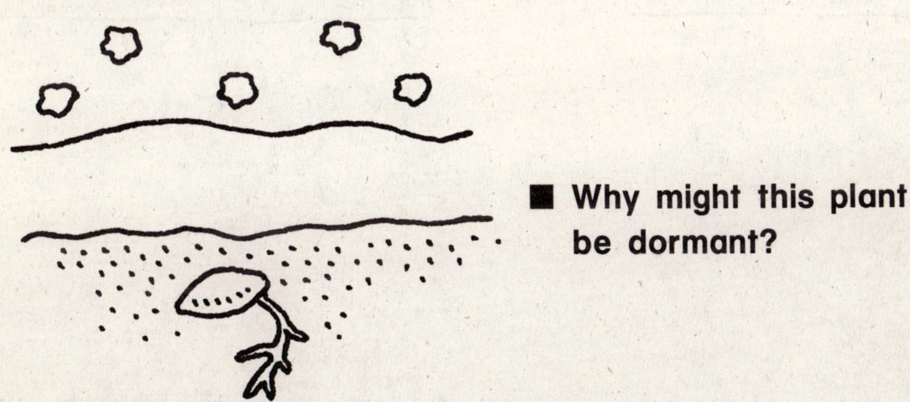

■ **Why might this plant be dormant?**

Water and Plants

Water also helps plants grow. When it rains, plants get the water they need. When it does not rain for a long time, plants grow more slowly. Plants that do not get enough water wilt, or become weak. If too much time passes without rain, the plants will dry up and die.

Name ______________________ Date __________

Activity

Directions How will the seed change? Draw what might happen.

Warm spring with lots of rain

Cold spring without much rain

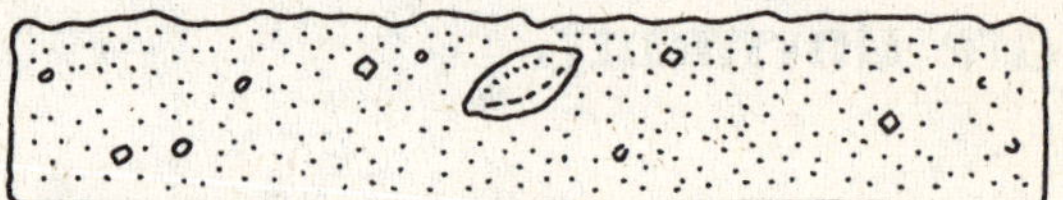

SOL 2.7a Use with Unit A, Chapter 1.

Name ______________________ Date ____________

LESSON 5

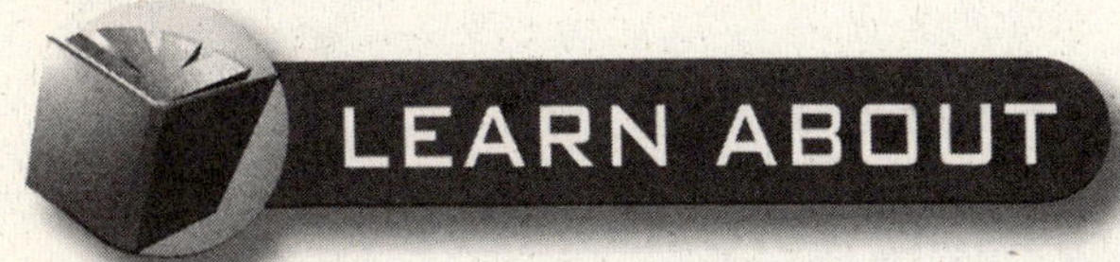

Vocabulary
gas **oxygen**

Plants Make Oxygen

Plants Are Important

Plants are important for many reasons. They provide homes and foods for many living things. Plants even help living things breathe.

Plants Make Oxygen

Gas is a kind of matter. The air around us is made up of different kinds of gas. One gas in the air is **oxygen.** Living things need to breathe oxygen.

Plants make oxygen. The sun shines on the leaves of green plants and gives the plants energy. The leaves use this energy to make food for the plant. As they make the food, they also make oxygen that goes into the air.

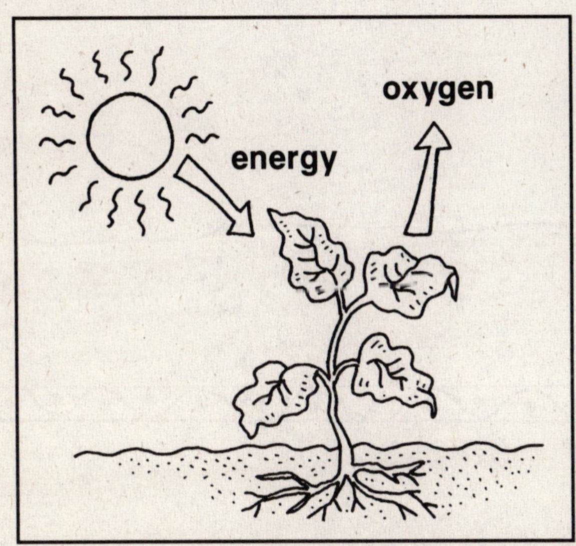

■ **What does sunlight help plants do?**

Name ______________________ Date ____________

Activity

Directions Draw a picture of a plant. Use green to color the part of the plant that makes food. Then write the word that tells what else that part of the plant makes.

Name ______________________ Date ____________

LESSON 6

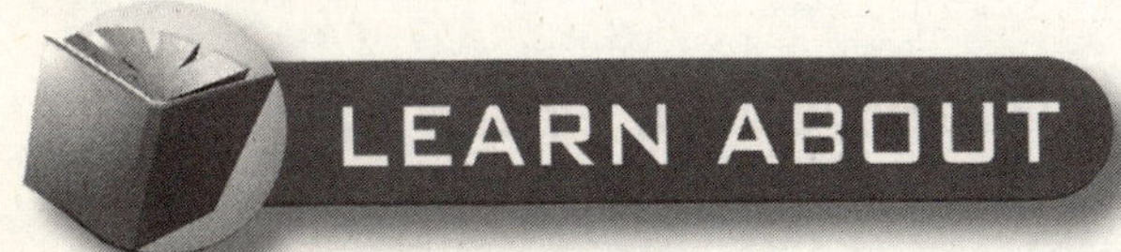

Vocabulary

interpretation

Observation and Interpretation

Observe and Draw Conclusions

When you make an observation, you use your five senses. You look at or listen to something. After an adult tells you that it is safe, you might smell, taste, or touch the thing. After you have made your observation, you need to explain what you have observed. When you do that, you are making an **interpretation.** For example, you might see an ice cube sitting in a dish. Ten minutes later you might see a puddle of water in the dish. Your interpretation of your observations might be that the ice cube melted.

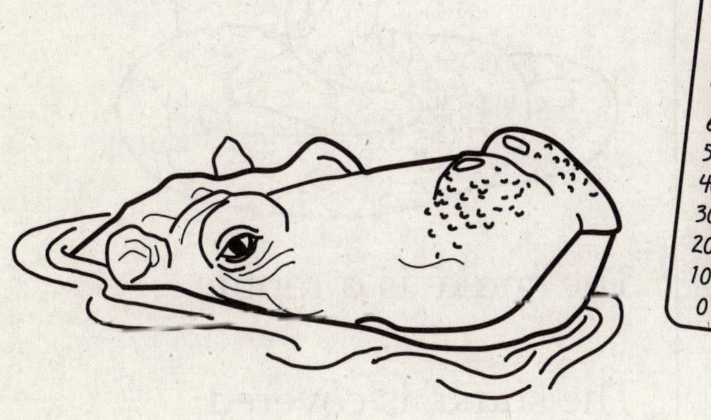

■ **Observe the hippo. Then give your observations.**

Name ______________________________ Date ____________

Activity

Directions For each picture, circle the observation. Underline the interpretation.

The turtle has a hard shell.

The turtle's shell protects it.

The polar bear can live where it is cold.

The polar bear has thick fur.

The monkey is swinging from a tree.

The monkey has lots of energy.

The snake is a reptile.

The snake is covered with scales.

Name ______________________ Date ____________

LESSON 7

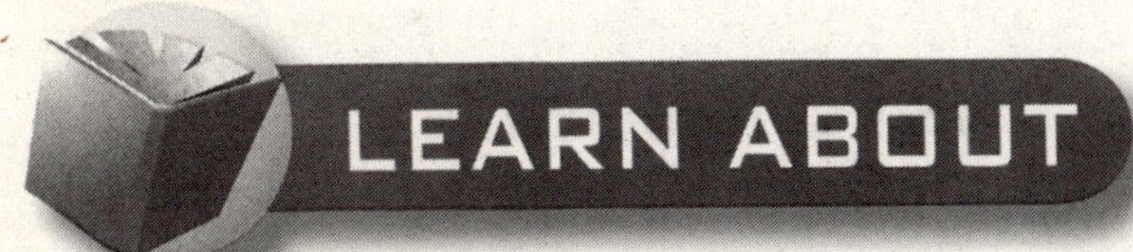

Vocabulary
life cycle

Animal Life Cycles

Animals Have Life Cycles

The **life cycle** of an animal is made up of all the parts of the animal's life. Animal life cycles begin in different ways. Most mammals grow inside their mothers' bodies. They look like their parents when they are born.

■ **Does this young elephant look like its mother?**

Animals That Hatch from Eggs

Other animals grow inside eggs. The mother lays the eggs. When the young animal is ready, it hatches, or breaks out of the egg. Many birds, insects, reptiles, and amphibians hatch from eggs. Some look very different from their parents. For example, tadpoles do not look like frogs.

When young animals become adults they are ready to have young of their own. The life cycle begins again.

■ **How does this snake's life cycle begin?**

Name ________________________________ Date ______________

Activity

Directions Look at the animal pictures. Then complete the chart.

Animal	Does this animal hatch?	Draw what it looks like as an adult.

Name ______________________ Date __________

LESSON 8

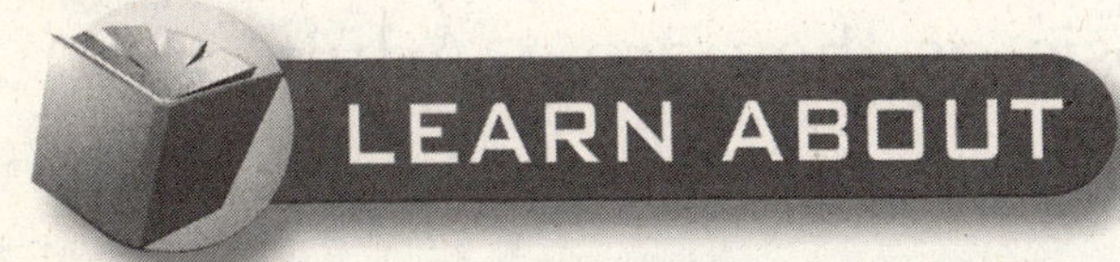

Vocabulary	
metamorphosis	**nymph**
larva	**pupa**

Metamorphosis

Animals and Their Parents

Some animals look like their parents from the beginning. Other animals look different from their parents at first. Some look only a little different, and some look very different. Many changes happen to these animals as they grow. With each change, they look more like their parents. This process is called **metamorphosis.**

Grasshoppers

A grasshopper begins its life cycle as an egg. When a young grasshopper hatches, it is called a **nymph.** It looks a lot like its parents, but it does not have wings.

As the grasshopper becomes an adult, it grows wings and can fly. A grown female is ready to lay its own eggs.

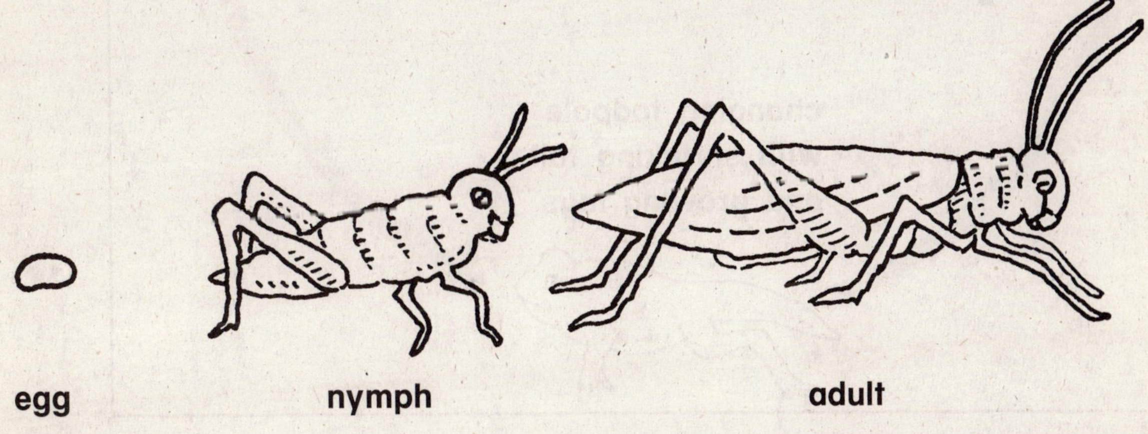

egg nymph adult

Name ________________________________ Date ______________

Frogs

Frogs go through metamorphosis as they grow. Adult frogs lay their eggs in water. Young frogs, or tadpoles, hatch from the eggs. Tadpoles do not look like adult frogs. They have tails for swimming and gills to breathe in water. They do not have legs. They look like tiny fish.

Tadpoles go through many changes. They grow legs. Their tails become smaller. They get lungs to breathe air. After a while, the growing frog can live on land.

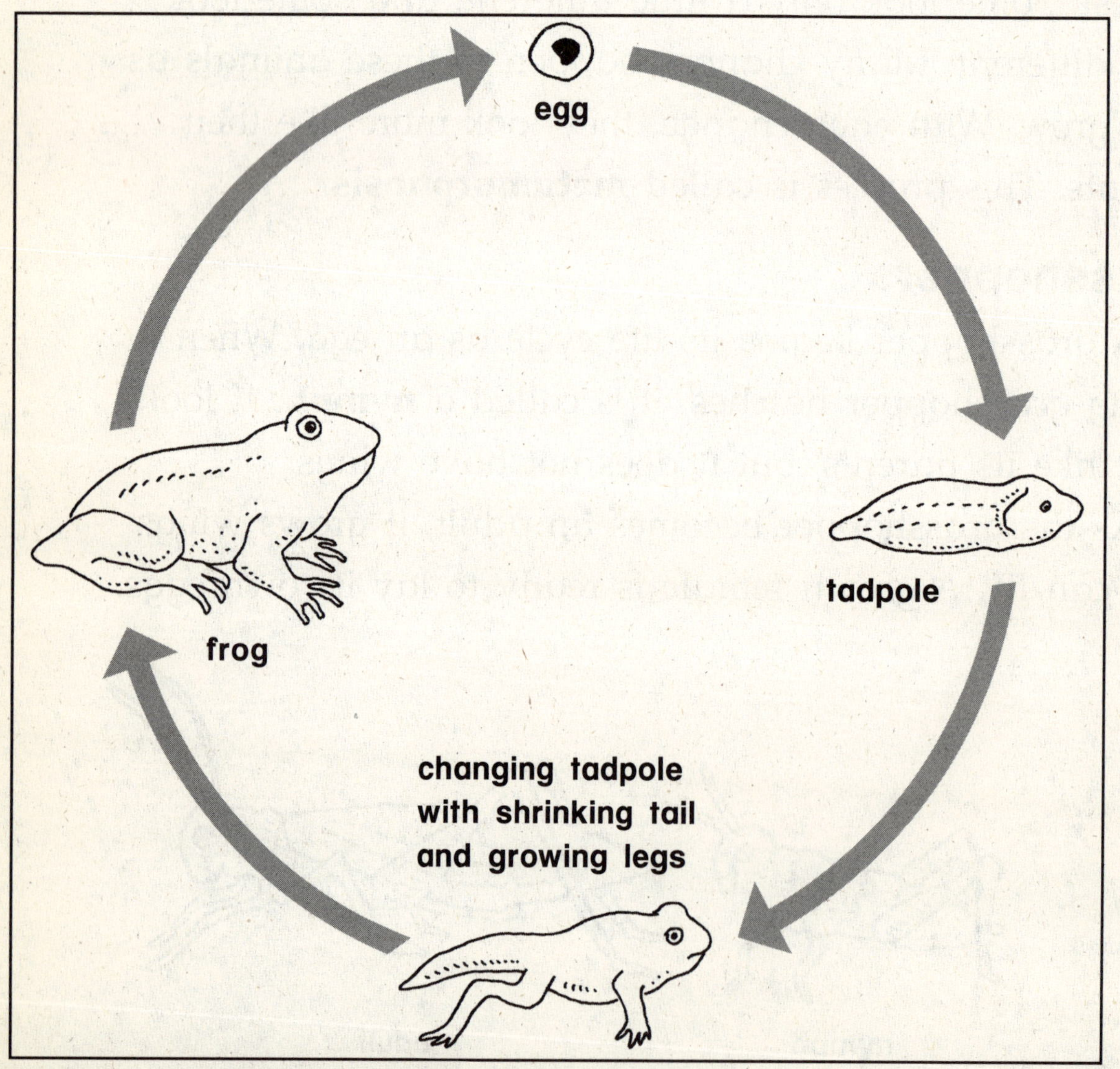

■ **How is the frog changing?**

Name ______________________ Date ____________

Butterflies

Butterflies also go through metamorphosis. Butterflies lay their eggs on plant leaves and stems. A caterpillar, or **larva,** hatches from each egg. After it hatches, it goes through many changes.

The caterpillar eats lots of leaves and grows. It sheds its skin as it grows. Then the caterpillar stops eating. It looks for a place to rest.

The caterpillar is ready for the next change. It makes a hard covering around its body and becomes a **pupa.**

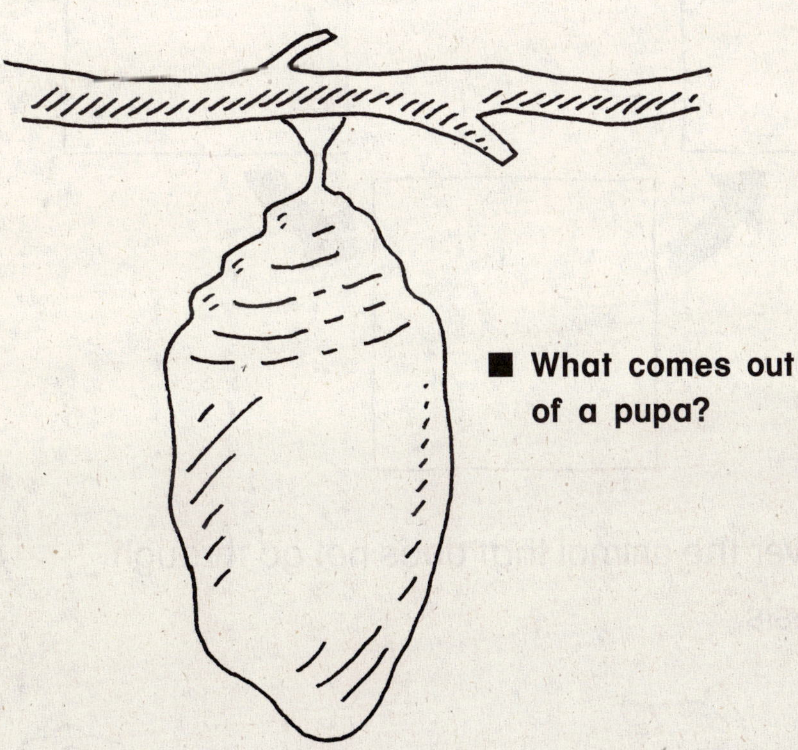

■ **What comes out of a pupa?**

Inside the covering, the pupa is changing into a butterfly. When it is ready, the covering opens. The butterfly unfolds its new wings and then flies away.

Name ______________________ Date ______________

Activity

Directions ❶ Draw the parts of a butterfly's life cycle.

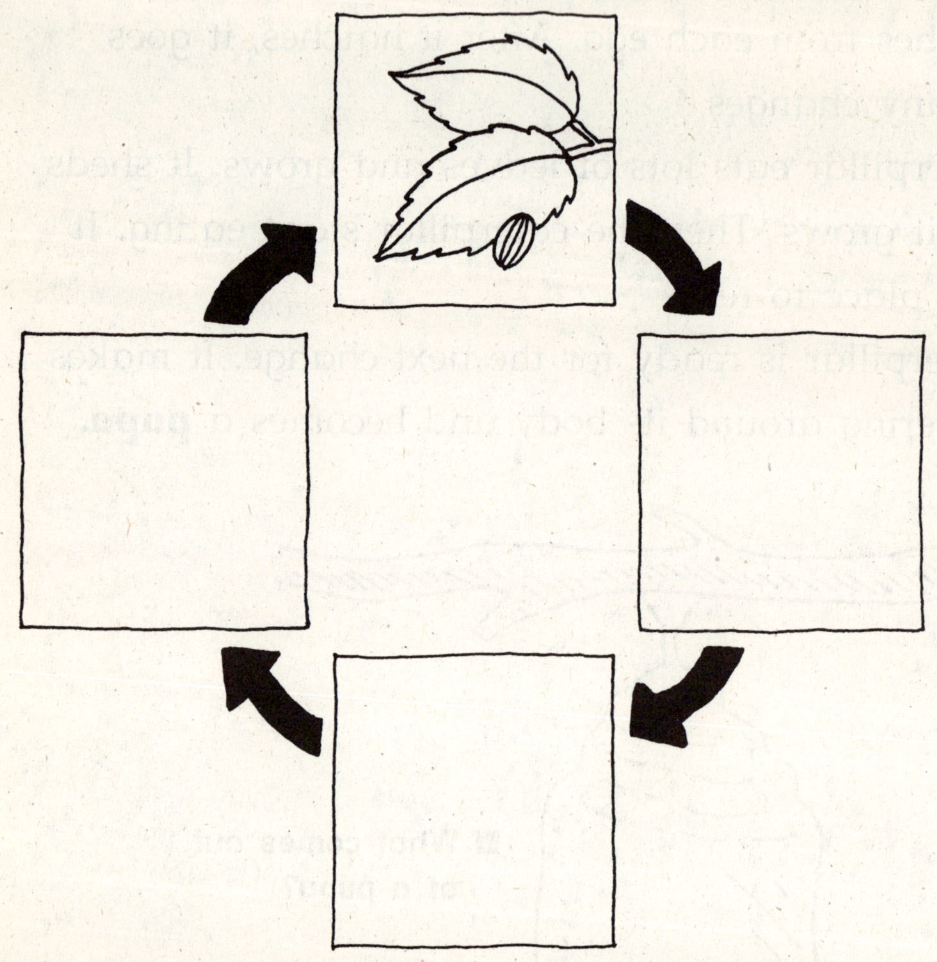

❷ Draw an *X* over the animal that does not go through metamorphosis.

Name ______________________ Date ____________

LESSON 9

Repeated Observations

You will need
- a picture
- a pencil

1. Work with a partner. Your partner will count while you observe the picture.
2. Have your partner quietly count to 20.
3. While your partner is counting, observe the picture. List on the worksheet the items you observe in the picture.
4. Repeat Step 3 two more times. You may write the same things again.
5. Trade jobs with your partner. Let your partner observe while you count.

Name ______________________________ Date ____________

Record Your Observations

What do you see in the picture? In the chart, write what you observe. Use words that describe.

First Time
Second Time
Third Time

Look at your chart. When did you observe the most things—the first, second, or third time?

____

How did looking again help you observe more?

____

Name ______________________ Date ____________

LESSON 10

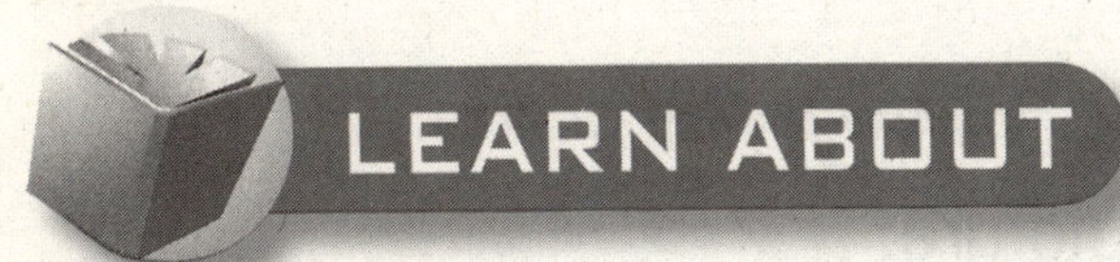

Vocabulary
environment **system**

Exploring a System

Living Things

Most living things need other living things to survive. Some animals eat plants, and some eat other animals.

Living things find the things they need in their environment. Their **environment** is made up of all the other living and nonliving things around them.

■ **What is this rabbit using from its environment?**

Living things also use nonliving things. Plants use water and sunlight to grow. Animals drink from ponds and lakes. Some animals live in water. Others hide under rocks or dig tunnels in the soil.

All the living and nonliving things in an environment work together. They make up a system. A **system** is a group of things that work together.

■ **How is this snake using its environment?**

Name ______________________________ Date ____________

Activity

Directions Draw a picture of a system. Show three or more things that work together. Label your picture.

Name ______________________ Date ____________

LESSON 11

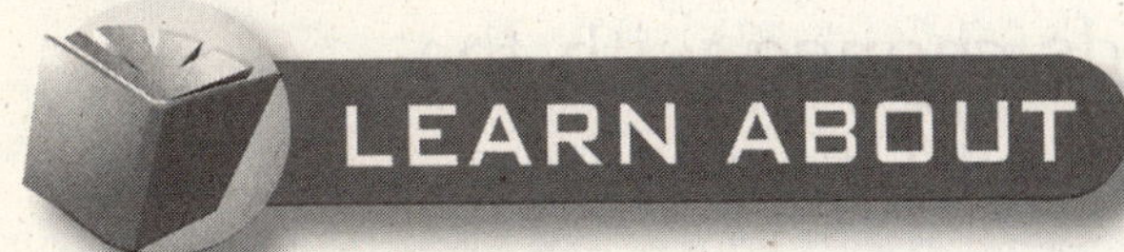

Vocabulary
habitat **grassland**

Changing Habitats

Forests

The place in which a plant or an animal lives is its **habitat.** Habitats change for many reasons. For example, forest habitats change with the seasons. In Virginia forests, the leaves of some trees change color with the change of seasons.

■ **How might this forest be different in winter?**

Grasslands

A **grassland** is a habitat made up of mostly grass and flat land. People have changed many grasslands. They have built farms and planted crops or built cities and towns.

■ **How have people changed this grassland?**

Name ______________________ Date ____________

Rivers, Streams, and Ponds

Many rivers, streams, and ponds change with the seasons. Cold winter temperatures can freeze rivers, streams, and ponds. Hot summers without rain can dry them up.

■ **How does this stream change in winter?**

People have also changed rivers, streams, and ponds. They have taken fish from them for food. Too much fishing changes a water habitat.

■ **How is this person changing this pond habitat?**

Name ______________________________ Date ______________

Activity

Directions This picture shows a grassland habitat. Draw things in the picture to show how people can change this habitat.

Name ______________________ Date __________

LESSON 12

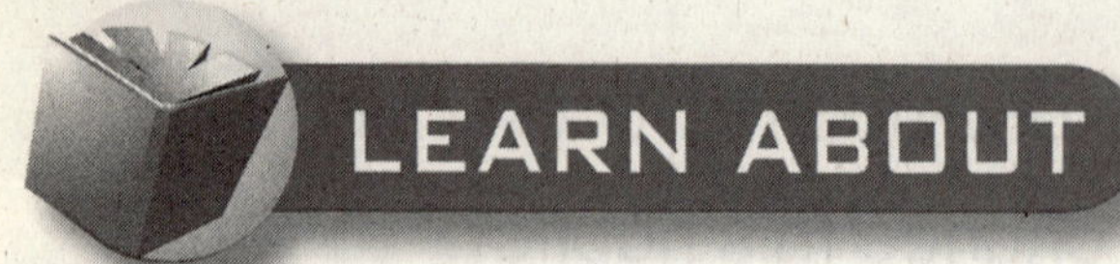

Vocabulary

weathering **erosion**

Weathering and Erosion

Earth Changes

The top layer of Earth is made of rock and soil. Tall mountains and cliffs have more rock. The land we use for planting and building has more soil.

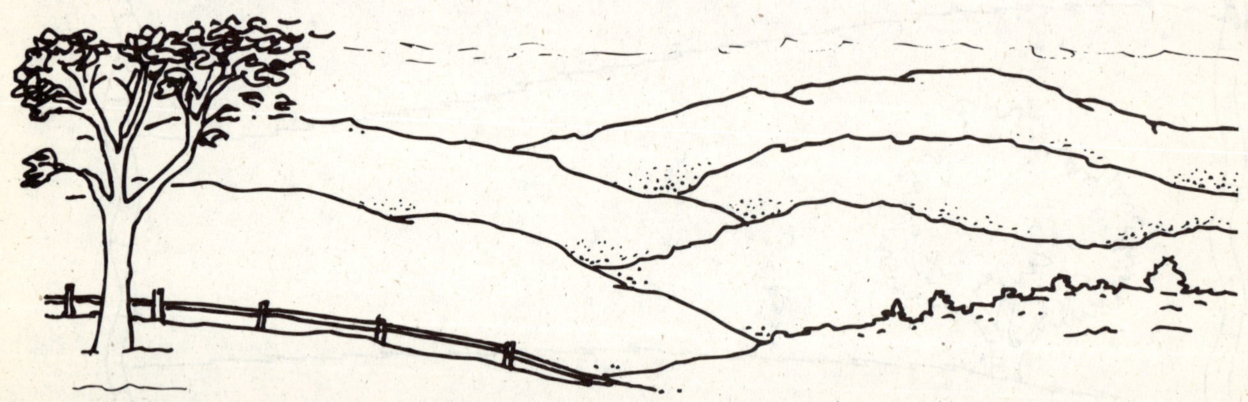

■ **What makes up the Blue Ridge Mountains in Virginia?**

Name ______________________ Date ____________

Weathering

Tree roots and ice break rocks into smaller pieces. Rushing water and sand blown by wind scrub rocks smooth. This process is called **weathering.** Over time, weathering changes the land.

Weathering happens in many ways. Water that flows over rocks may carry sand and smaller rocks. The sand and rocks work like sandpaper to wear down big rocks. Rain gets into the cracks of big rocks. When the weather gets very cold, the water freezes inside the cracks. This makes the rocks break.

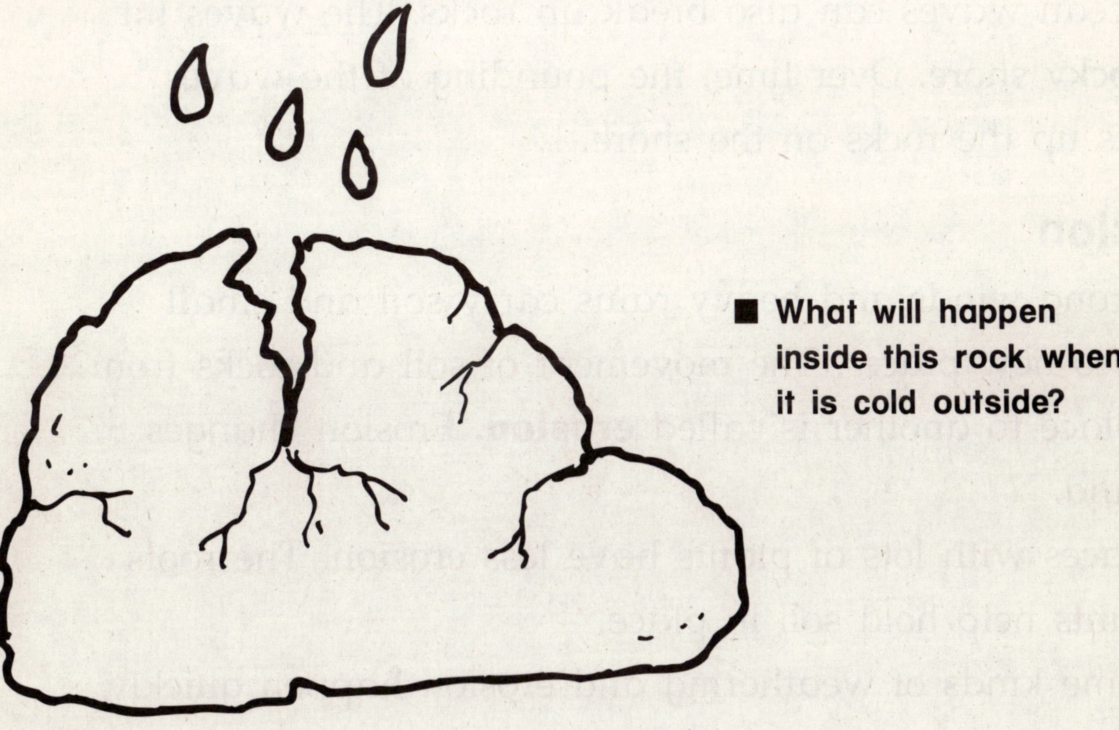

■ **What will happen inside this rock when it is cold outside?**

Name ______________________ Date __________

Rocks are also broken up by trees. Tree roots grow into cracks in the rock. When the roots grow larger, they make the rocks break.

■ **What is breaking up these rocks?**

Ocean waves can also break up rocks. The waves hit the rocky shore. Over time, the pounding of the waves breaks up the rocks on the shore.

Erosion

Strong winds and heavy rains carry soil and small rocks to new places. The movement of soil and rocks from one place to another is called **erosion.** Erosion changes the land.

Places with lots of plants have less erosion. The roots of plants help hold soil in place.

Some kinds of weathering and erosion happen quickly, but most kinds take a long time.

Name ______________________ Date ____________

Activity

Directions 1 Put the pictures in order to show how weathering happens.

____

1

____

____

2 Complete the sentences.

______________________ breaks up rocks into smaller pieces.

______________________ moves rocks and soil to new places.

Erosion **Weathering**

Name ______________________ Date __________

LESSON 13

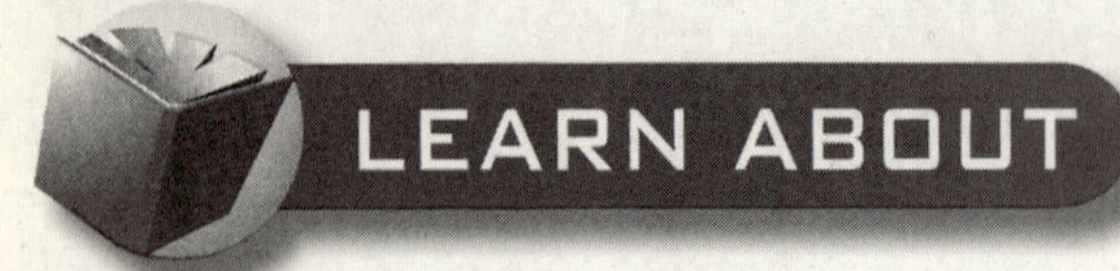

Vocabulary

cotton **linen** **lumber**

Uses of Plants

Cloth from Plants

Many things that people use are made from plants. **Cotton** comes from a plant. After the plant has flowers and makes seeds, the cotton forms. It can be used to make thread and then cotton cloth.

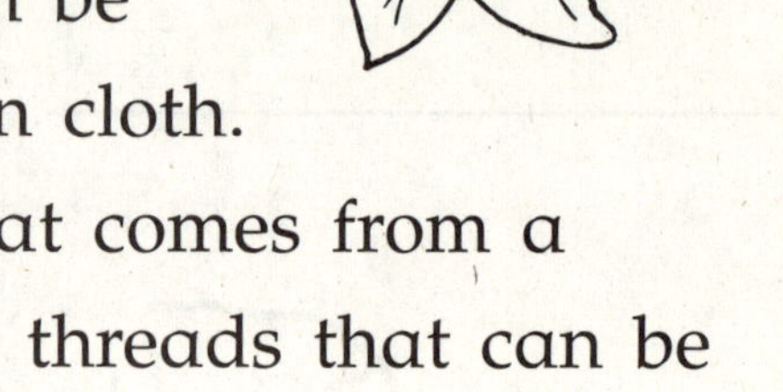

Linen is another kind of cloth that comes from a plant. Inside the stem there are long threads that can be made into linen cloth.

Other plants also have parts that can be made into thread, string, and rope.

The cloth that comes from plants can be used to make clothing. Jeans can be made from cotton cloth. Some dresses are made from linen cloth.

■ **What are these jeans made from?**

Lumber from Trees

Trees provide **lumber,** or wood that is cut into pieces for building and making things. Many people use lumber for building houses and making furniture.

Wood is a good building material because it is strong and lasts a long time.

Name ______________________ Date __________

Activity

Directions 1 Circle the objects that are made from plants.

2 Look around your classroom. Find four things made from wood and four things not made from wood. Then draw or write to complete the chart.

Things Made from Wood	Things Not Made from Wood

Name ______________________ Date ____________

LESSON 14

Vocabulary
crops

Where Plants Grow

Plants, Temperature, and Soil

Plants do not grow in all places. They need certain temperatures to grow. They also need certain kinds of soil. Banana trees need warm temperatures all year round. Cacti need dry, sandy soil.

People and Plants

Most people live in places where plants grow. They use some plants for food and other plants for building houses and furniture. When people go to a place without plants, such as a desert, they must take along these things to survive.

Farmers in Virginia grow many crops. **Crops** are plants grown to provide foods and other materials. Apples, peaches, and peanuts are important crops in Virginia. Peanuts are used to make many foods.

peanut plant

Trees from Virginia's forests are used to make paper, furniture, and houses. Some of the trees that are used are pine, oak, and hickory trees.

hickory tree

Name ______________________ Date ____________

Activity

Directions The plants listed below grow in Virginia. Choose two, and write their names in the chart. Then draw a picture to show how people use each plant.

Peanut Apple Peaches Oak

Name of Plant	How People Use the Plant

Name ______________________ Date __________

LESSON 15

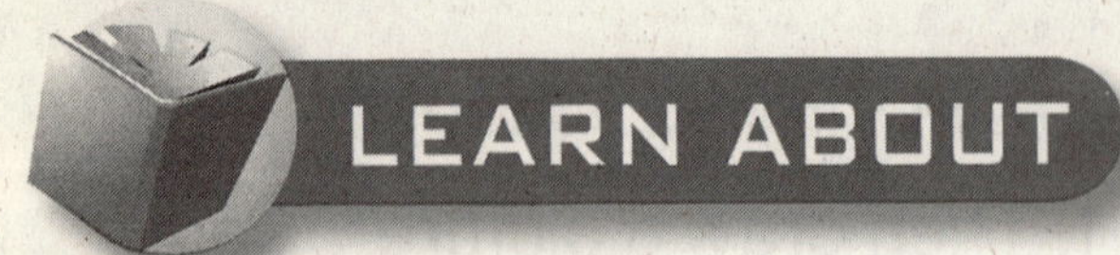

Vocabulary

snow **sleet**

Winter Weather

Too Cold for Rain

When it rains, drops of water fall from the clouds. In winter the water may freeze in the clouds or as it falls. Then snow or sleet falls instead of rain.

Snow forms when the water in a cloud freezes into snowflakes. Snowflakes are tiny ice crystals with beautiful shapes.

■ **How many sides does this snowflake have?**

Sometimes drops of rain fall from a cloud and then pass through cold air. They freeze as they fall and form **sleet.** Sleet looks like tiny balls of ice.

Activity

Directions Label each picture.

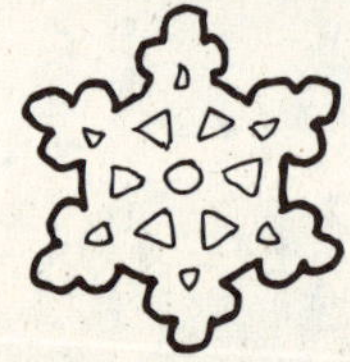

_______________ _______________

Name ______________________ Date __________

LESSON 16

Vocabulary
data **predict**

Interpreting Data

Collecting Data

Scientists collect information, or **data,** when they study something. They show the data in tables, charts, and graphs.

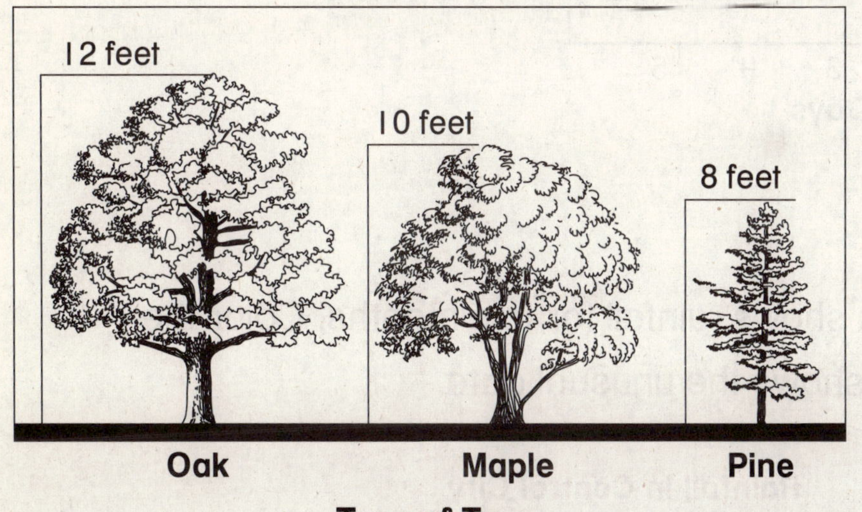

■ **What does the diagram show?**

Scientists **predict** what will happen before they do an experiment. They use what they already know to make a good guess, or a prediction.

Name ________________________ Date ____________

An experiment gives scientists data. Sometimes the data is very different from their prediction. Scientists work to explain this different, or unusual, data.

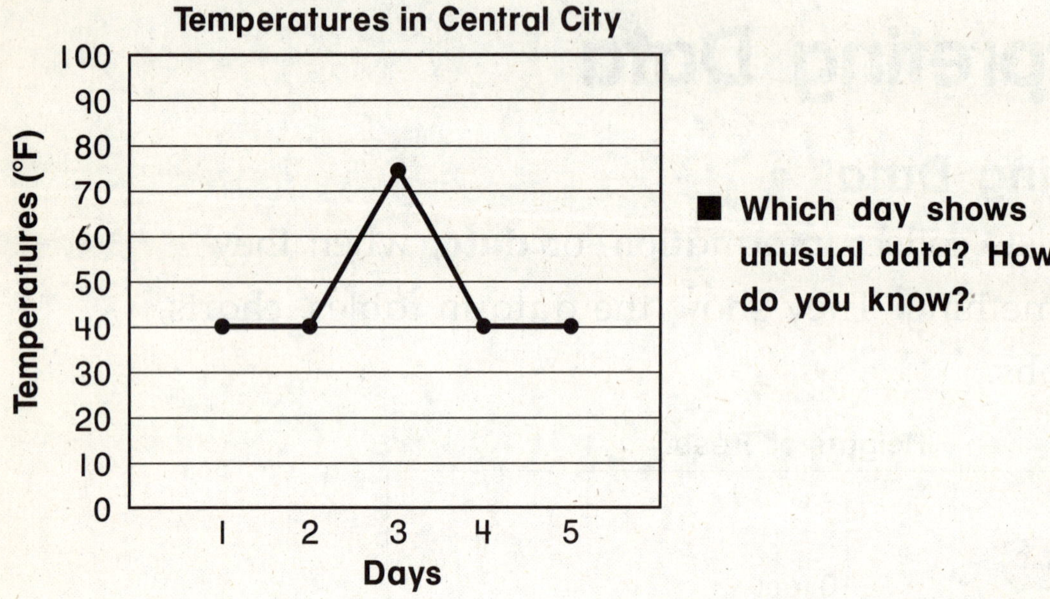

■ **Which day shows unusual data? How do you know?**

Activity

Directions The graph shows rainfall for five months. Color the part of the graph that shows the unusual data.

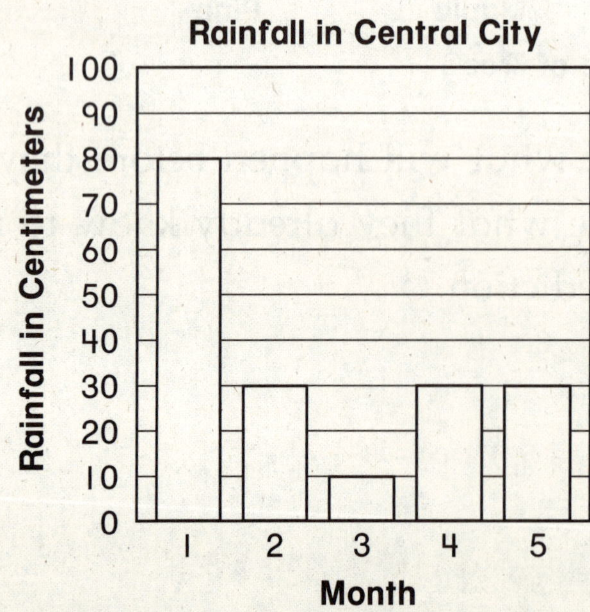

Name ______________________ Date ____________

LESSON 17

Vocabulary
metric units
customary units
mass volume

Using Measurement

Using Different Units to Measure

Some people use metric units to measure things. **Metric units** include centimeters, meters, grams, kilograms, milliliters, liters, and degrees Celsius. Most scientists use metric units.

In the United States, people also use customary units. **Customary units** include inches, feet, yards, ounces, pounds, cups, pints, gallons, and degrees Fahrenheit.

The chart below shows some metric units and customary units.

What Is Measured	Metric Units	Customary Units
length	centimeters meters	inches feet yards
mass	grams kilograms	ounces pounds
volume	milliliters liters	fluid ounces cups pints quarts gallons
temperature	degrees Celsius	degrees Fahrenheit

Name ______________________ Date __________

Measuring Length

You can use a ruler to find out how long an object is. Some rulers show metric units. You can measure a short object in centimeters and a longer object in meters. Some rulers show customary units. You can measure objects in inches, feet, or yards.

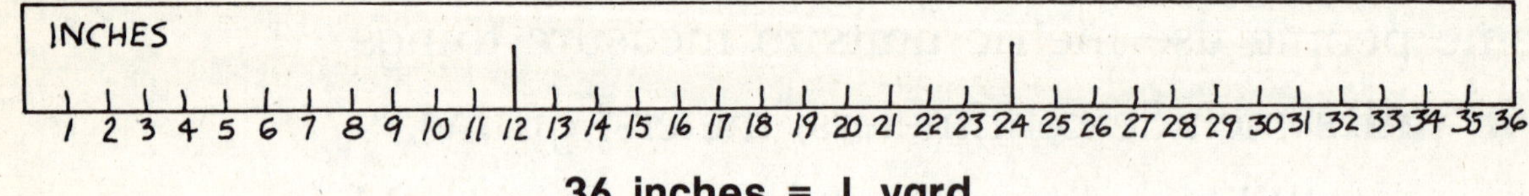

36 inches = 1 yard

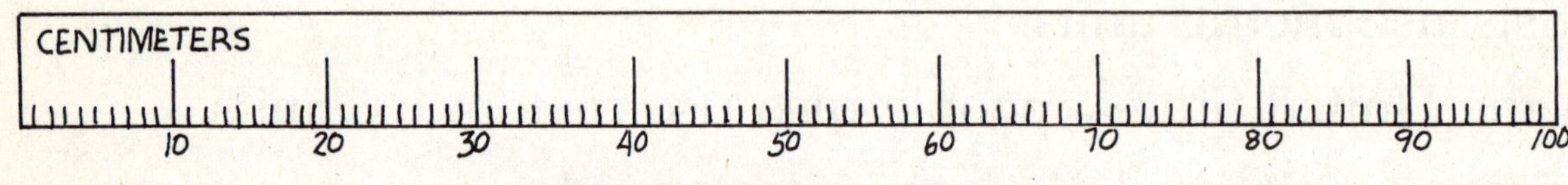

100 centimeters = 1 meter

Measuring Mass

Grams and kilograms are metric units used to measure mass. **Mass** is the amount of matter an object has. An object with a small mass is measured in grams. An object with a larger mass is measured in kilograms.

In customary units, an object with a small mass is measured in ounces. An object with a larger mass is measured in pounds.

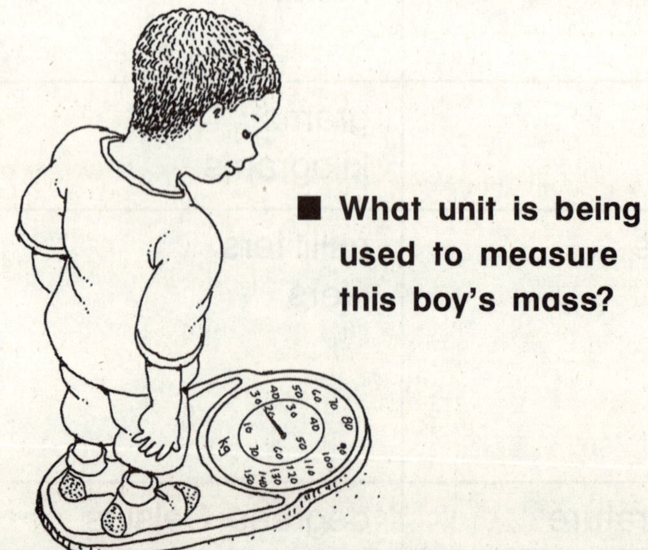

■ **What unit is being used to measure this boy's mass?**

Name ______________________ Date __________

Measuring Volume

Liters and milliliters are metric units used to measure volume. **Volume** is the amount of space that something takes up. A liquid with a small volume is measured in milliliters. A liquid with a larger volume is measured in liters.

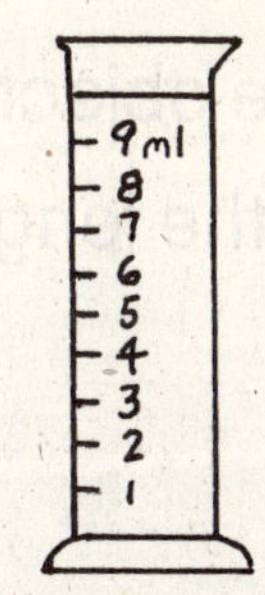

■ **What metric units are used to measure volume?**

In customary units, cups and pints are used to measure small volumes of liquids. Quarts and gallons are used to measure larger volumes. Many things we drink are measured this way.

■ **What unit was used to measure the volume of this milk?**

Measuring Temperature

Temperature is the measure of how hot or cold something is. Thermometers measure temperature in units called degrees. The metric units used to measure temperature are degrees Celsius.

The customary units used to measure temperature are degrees Fahrenheit. In the United States, weather reports give the temperature in degrees Fahrenheit.

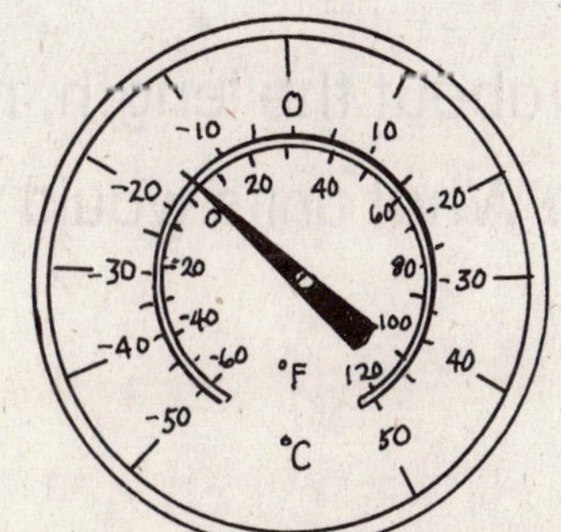

■ **What is the temperature on this thermometer?**

Name ______________________________ Date ____________

Activity

Directions 1 Find three objects in your classroom. Write their names in the chart. Find the length of each object in centimeters and in inches.

Object 1 ____

________________ centimeters

________________ inches

Object 2 ____

________________ centimeters

________________ inches

Object 3 ____

________________ centimeters

________________ inches

2 Write some questions about the length, mass, volume, or temperature of things. What units would you use to answer each question?

Name ____________________ Date __________

LESSON 18

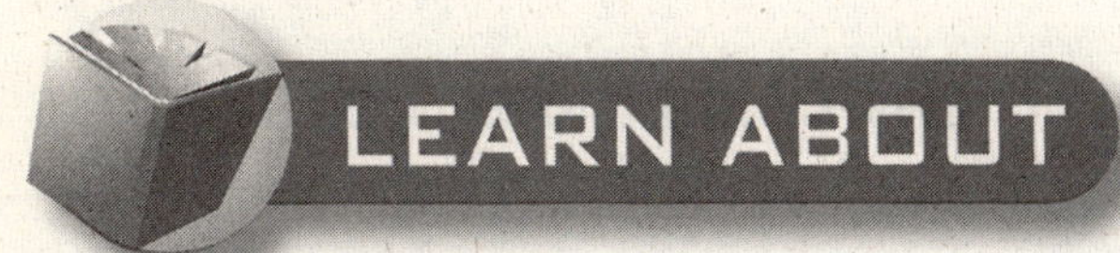

Vocabulary
graphs

Labeling Graphs

Graphs Show Data

One way scientists show data is by using graphs. **Graphs** are a good way to show others the data you collect in an experiment.

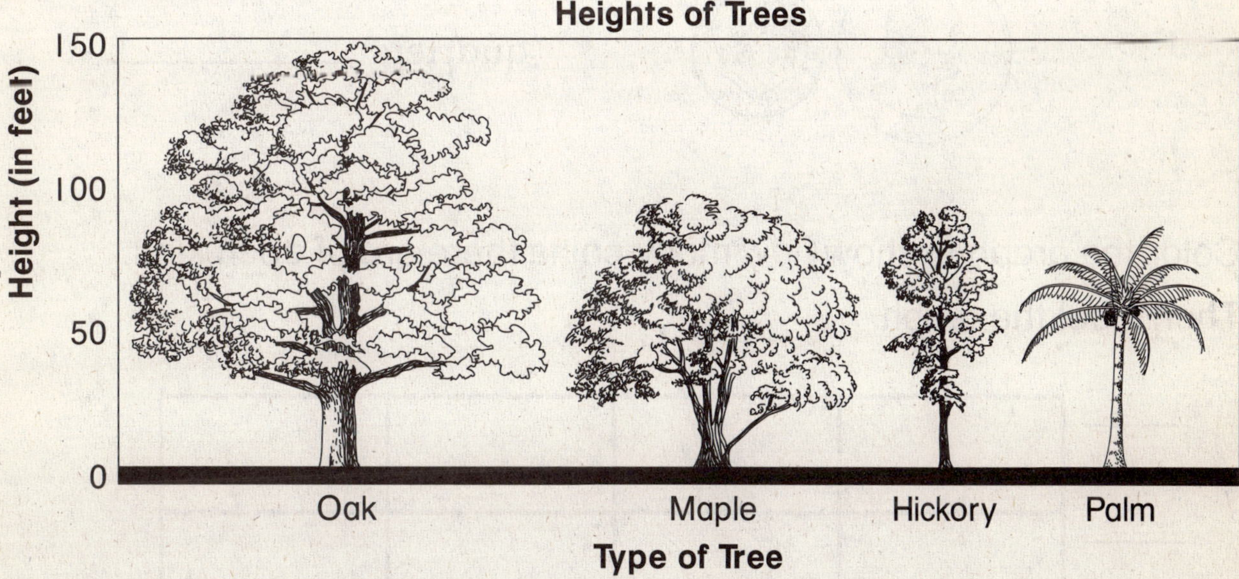

■ **What can you tell from the graph?**

Graphs need labels to explain what the data shows.

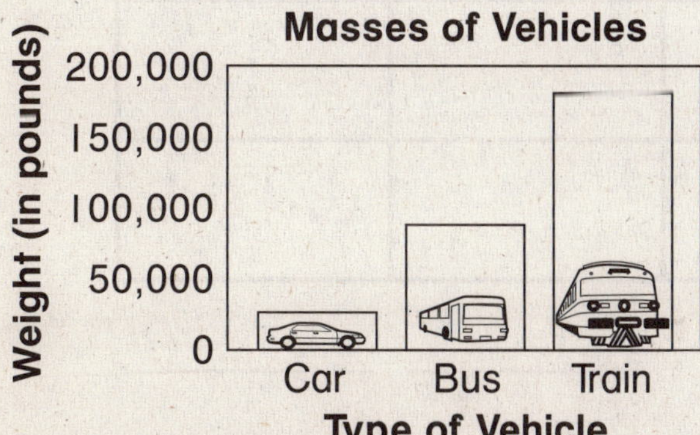

■ **What are the labels on this graph?**

Name ______________________ Date ____________

Activity

Directions Count the coins. Write how many coins there are of each kind.

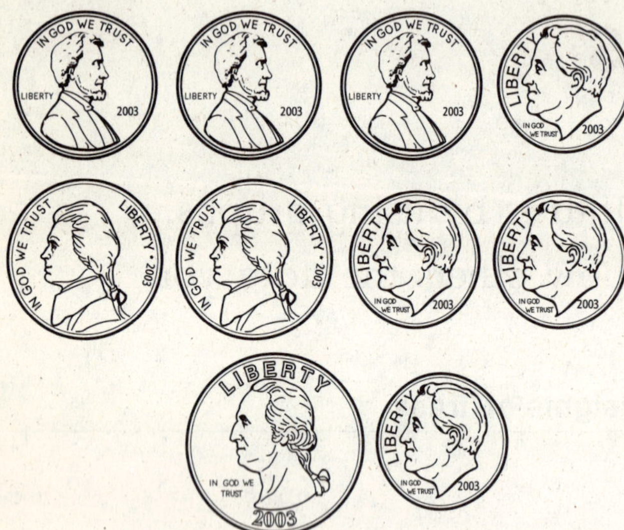

pennies ________

dimes ________

nickels ________

quarters ________

Color the graph to show how many coins there are of each kind. Then label the graph.

____				
____				
____				
2				
1				
0	Pennies	Nickels	________	________

Name ______________________ Date __________

LESSON 19

Vocabulary

milliliter **fluid ounce**

Volume

Matter Has Volume

Volume is the amount of space that matter takes up. Solids, liquids, and gases all take up space. This means that all matter has volume.

Solids

When you look at a solid, you can see that it takes up space. Solids that take up more space have more volume.

■ **Which ball has more volume?**

Name ______________________ Date __________

Liquids

You know that milliliters, liters, cups, pints, quarts, and gallons all measure the volume of liquid. These units tell how much space a liquid takes up.

Milliliters are the metric unit for measuring small volumes of liquids. A **milliliter** is the amount of liquid in about 20 drops of water. Fluid ounces are the customary unit for measuring small volumes of liquids. A **fluid ounce** is about two big spoonfuls of a liquid.

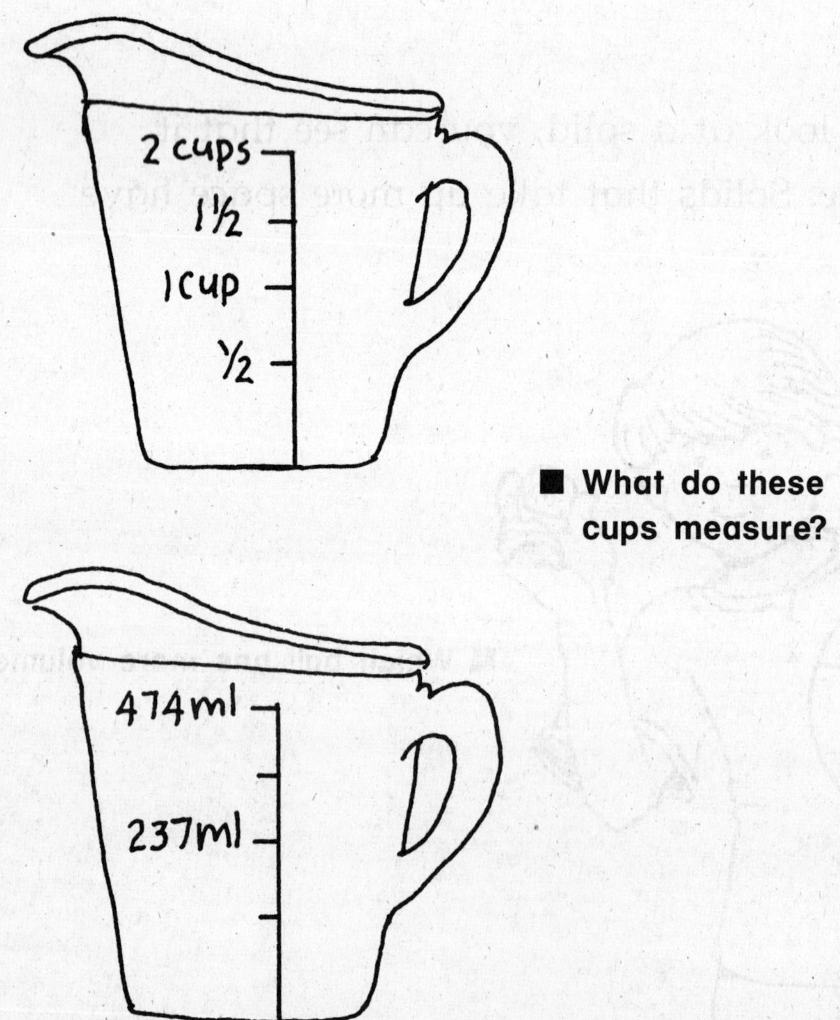

■ **What do these cups measure?**

Name ______________________ Date ____________

Gases

Remember that a gas is the only kind of matter that always fills all the space inside a container. Gas takes up space, so it has volume. You can look at the size of the container to see its volume.

■ **How do you know the gas inside the balloon has volume?**

"Empty" containers are not really empty. They are filled with air, which is a gas. Larger containers hold more gas. The gas inside them has more volume.

■ **Which cup holds more gas? How do you know?**

Name ______________________ Date ____________

Activity

Directions Look for things in your classroom that show each kind of matter. List them where they belong in the chart. Tell how you could measure the volume of each one.

Kind of Matter	How to Measure Its Volume
Solid	
Liquid	
Gas	

Name ______________________ Date __________

LESSON 20

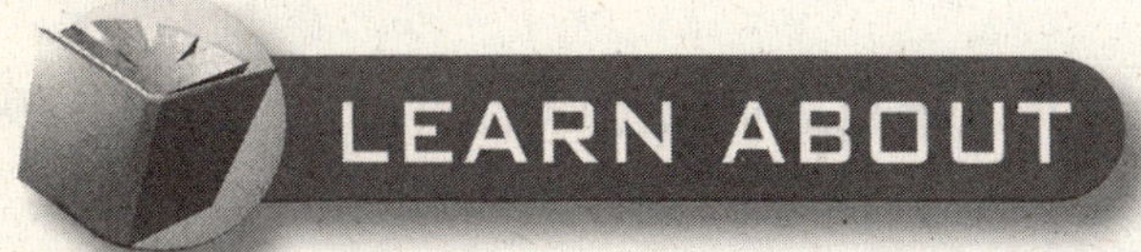

Vocabulary
physical change

Physical Changes

Matter Changes Form

Have you ever seen people ice-skating on an outdoor pond or lake in winter? Winter is a time of freezing temperatures.

The water in ponds and lakes freezes into ice when the temperature is very cold. When the temperature becomes warmer, the ice melts back into water. It is still the same water whether it is frozen or not.

Matter has three states—solid, liquid, and gas. A change from one state to another is called a **physical change.** Some solids can change into liquids. Some liquids can change into solids.

Heating an ice cube makes the solid form of water change into a liquid. Freezing the water makes the liquid form of water change into a solid.

■ **How is the ice cream changing?**

Name ______________________ Date ____________

Activity

Directions Write a sentence to describe the physical change that is happening in each picture.

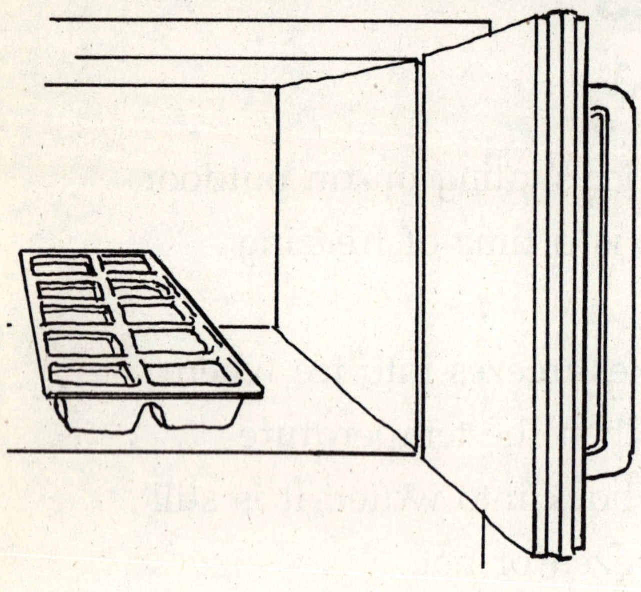

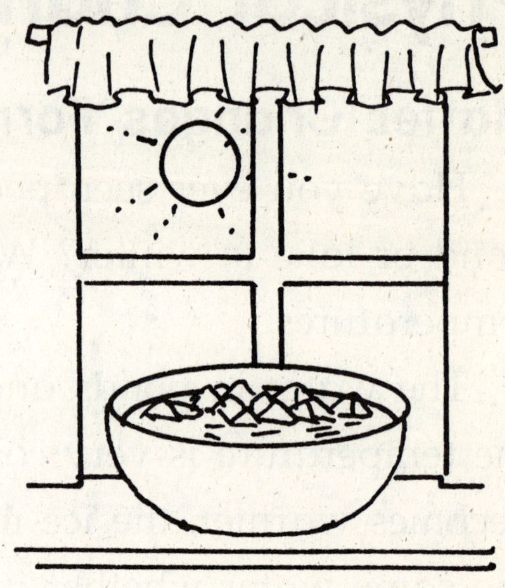

______________________ ______________________

______________________ ______________________

______________________ ______________________

Name ______________________ Date ____________

LESSON 21

Vocabulary

magnetite **nickel**

Properties of Magnets

Magnets Attract Certain Things

Some magnets are found in nature. Lodestone is a kind of rock that is magnetic. Lodestone is also called **magnetite.**

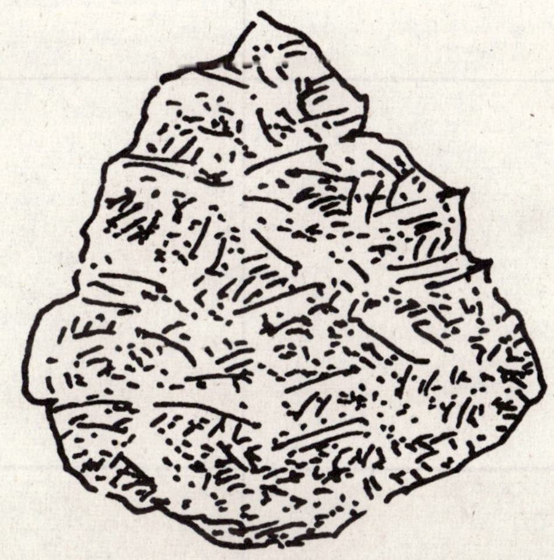

magnetite

Magnets attract only certain kinds of objects. They attract objects made of iron or steel. They also attract objects made of nickel. **Nickel** is a kind of metal.

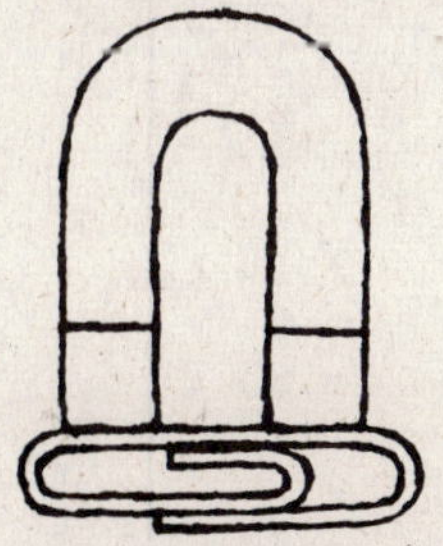

steel paper clip

Name ______________________________ Date ____________

Activity

Directions Predict what will happen when you hold a magnet near each object. Then test your prediction.

Object	*Will* a magnet attract it?	*Did* the magnet attract it?
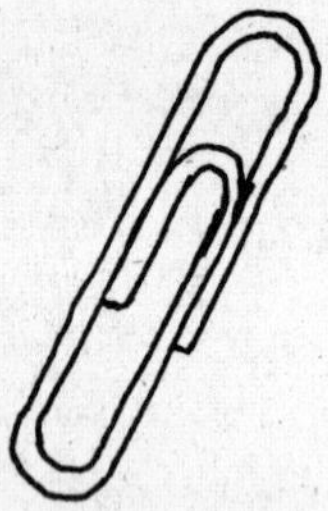		

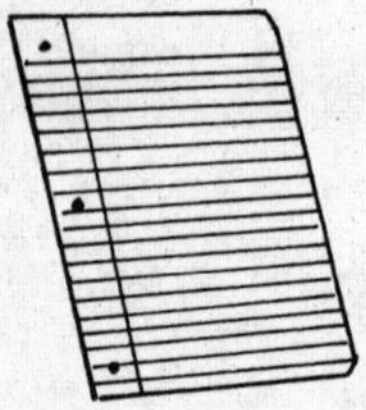		

Name ______________________________ Date ______________

LESSON 22

INVESTIGATE

Magnetic Compasses

You will need
- a dull needle
- a bar magnet
- a flat piece of cork
- tape
- a small bowl of water
- chart

Directions

1. Rub the needle with the north end of the magnet. Rub in only one direction. Do this 20 times.
2. Tape the center of the needle to the piece of cork.
3. Place the cork in the bowl of water.
4. Is the point of the needle facing north, south, east, or west? How do you know?
5. In the chart, draw what you **observe** when you face north. **Infer** where south, east, and west are. Draw what you see when you face each way.

Name ______________________ Date ____________

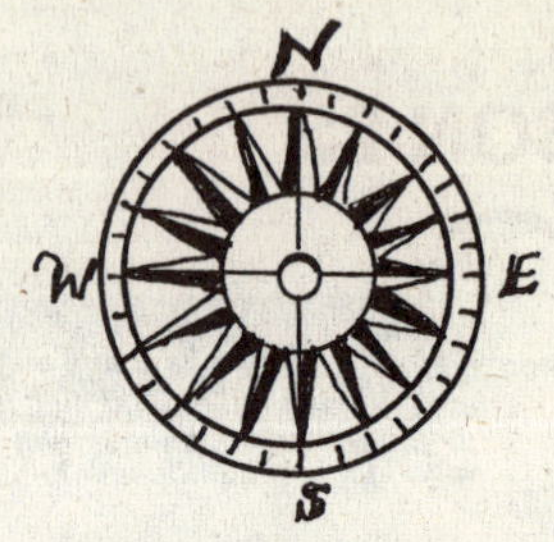

Record Your Observations

One thing I see when I look north	One thing I see when I look east
One thing I see when I look south	**One thing I see when I look west**

Science Skill You can make a compass and **observe** how it works.